Life and Worl

By Shalu Sharma

Aki
Akshat Gode

205-401-8347

Copyright:

Copyright 2013 Shalu Sharma. All Rights Reserved. http://www.shalusharma.com

No portion of this book may be modified, reproduced or distributed mechanically, electronically, or by any other means including photocopying without the written permission of the author.

Disclaimer:

Whilst every care is taken to ensure that the information in this book is as up-to-date and accurate as possible, no responsibility can be taken by the author for any errors or omissions contained herein.

DEDICATED TO ARYABHATA

Table of contents

Who was Aryabhata? 4

World's greatest mathematicians 10

Indian mathematicians 14

Ancient Indian mathematics 17

Indian mathematics 20

Introduction to Aryabhata 23

Name and birth place of Aryabhata 27

Taregna – The (birth) place of Aryabhata 32

The works of Aryabhata 36

The Arya-Siddhanta 39

Who invented Pi? 43

Approximation of Pi by others and Aryabhata 47

Aryabhata was not the first to use zero 50

The real story of zero 53

History of algebra 58

Aryabhata and algebra 62

Aryabhata and trigonometry 66

Indian astronomy and Aryabhata 69

Astronomical observations of Aryabhata 73

Heliocentrism and Aryabhata 77

References and further reading 80

Who was Aryabhata?

Aryabhata (476–550 A.D.) was an ancient Indian mathematician and astronomer who lived in India during the classical period of the Gupta Dynasty. When he was just 23 years old, he wrote his mathematical treatise called the '**Aryabhatiya**' in 499 A.D. It is thought that he was born on the 21st of March in the year 476 A.D. He has been named as Aryabhata I to distinguish him from another Aryabhata II, the author of Maha-Siddhanta who lived around 1000 A.D.

Aryabhata has been mentioned in Arab texts and was referred as 'Arjehir'. Not much is known about Aryabhata. We don't even know where he came from. Some historians say that he was from Kerala; some say that he was from Bihar and some say that he was from Central India and they apply their own logic to it. Some say that he was from the regions of the rivers of Narmada and Godavari situated in the Deccan most probably Ashmaka (South India). One commentator who commented on 'Arthashastra' the ancient Indian treatise on politics and military strategy written by 'Kautilya' has mentioned that Aryabhata was from Maharashtra.

My belief is that he was from a small town called '**Taregna**' just a few kilometres away from Patna, capital of Bihar. Taregna has been confirmed by NASA as one of the best places to watch solar eclipses. There has to be links as to why this place is considered a good place for such observations and Aryabhata being born there or at least living here and doing his observations. There is a small temple building, something of a shrine with astronomical connections in the town and people of the town say that it has been there for centuries. The name of the town literally means 'singing stars'.

The fact is that, Aryabhata lived and worked in Bihar and was the head of astronomy and mathematics at the ancient Nalanda University.

There is a lot of speculation about his name too. His name 'Arya Bhata' appears more like a North Indian name. Let's take a look at his name. 'Arya' or 'Aryan' is a typical North Indian boy's name and the surname 'Bhata' or 'Bhatta' (with a double t) is similar to 'Bhat' again is a typical North Indian surname. Unless his first name was Aryabhata combined together and had a separate surname which is unaccounted for, we must assume that his name was 'Arya' and 'Bhata' or 'Bhatta' which have been combined together either by himself or by others.

His treatise 'Aryabhatiya' mentions **'Kusumpura'** which was later known as Patliputra and now stands as modern day Patna. Kusumpura was the name of the fort that was established by King Ajatashatru in the third century, the king of Magadha at the confluence of four rivers which was initially known as "Kusumpura".

His words translate to;

"Having bowed with reverence to Brahma, Earth, Moon, Mercury, Venus, Sun, Mars, Jupiter, Saturn and the asterisms, Aryabhata sets forth here the knowledge honoured at Kusumapura".

> *praṇipatyaikaṃ anekaṃ kaṃ satyāṃ devatāṃ paraṃ brahma /*
> *Āryabhaṭas trīṇi gadati gaṇitaṃ kālakriyāṃ golam //*
> (Āryabhaṭīya, Gītikāpāda, verse 1)
>
> *brahma-ku-śaśi-budha-bhṛgu-*
> *kuja-guru-koṇa-bhagaṇān namaskṛtya /*
> *Āryabhaṭas tviha nigadati*
> *Kusumapure 'bhyarcitaṃ jñānam //*
> (Gaṇitapāda, verse 1)
>
> *Āryabhaṭīyaṃ nāmnā*
> *pūrvaṃ svāyambhuvaṃ sadā satyam /*
> *sukṛtāyuṣoḥ praṇāśaṃ*
> *kurute pratikañcukaṃ yo 'sya //*
> (Golapāda, verse 50)

Another noteworthy point is that he wrote in Sanskrit which was the language of the educated elite in the Northern plains of India. Whatever the case maybe, it is known that he lived in Nalanda and that he was the head (kulapati) of a faculty or of the Nalanda University where he taught mathematics. In addition, most authors and historians do acknowledge that he wrote his works in Nalanda during the kingdom of Buddhagupta (477 and 497 A.D.), the last ruler of the Gupta Dynasty. The time of the Gupta Dynasty is considered as the 'Golden Age' in Indian learning.

The kingdom of the Gupta Dynasty was Magadha (Bihar) with Patliputra (Patna) as the capital. It is known that Nalanda University was a great seat of learning and is known that Aryabhata was also the head of the astronomical observatory which was regarded in high esteem in those days. He was known as 'Kulapati' of the observatory. The word Kulapati simply means the 'head' similar to the vice-chancellor of a university. It is here at Nalanda University that he wrote the Arya-Siddhanta. Large parts of Arya-Siddhanta are lost. But we do know that it's a system of astronomy that was founded by none other than Aryabhata himself which could have been based on the Surya-Siddhanta which is thought to be one of earliest traditions (Siddhanta) in the astronomy of India.

In this book, we will look at the life history of this great Indian mathematician and find out the works he carried out. We shall also look at some of the other works of other mathematicians and how it compares to that of Aryabhata.

Note: Please note that the book does not detail his mathematical treatises and how he arrived at his mathematical or astronomical conclusions. We are exploring the life of this great Indian mathematician.

World's greatest mathematicians

The one language that every person on this planet has in common is Mathematics. As you may know, Mathematics is absolutely vital if we want to understand this planet. In this chapter, I want to take a little look at some of the greatest mathematicians in the world. This is a list of people who have shaped the way in which we view numbers today.

Aryabhata is the first in the list. This fantastic Indian Mathematician is credited with a number of great discoveries in the world of Mathematics. Most of these discoveries you have seen earlier on in this book. His major discoveries include an approximation of Pi, and he even helped towards the development of the number zero. I do not wish to dwell on his works for too long as I have gone into a lot of depth elsewhere in this book. Suffice to say though, without Aryabhata much of what we know about Maths would simply not exist.

Pythagoras! You all know him right? Well, it was this genius that was responsible for starting up study groups to look into advanced mathematics.

Much of what he discovered is still in use today. Perhaps one of the most important concepts that have been attributed to him is that of the Pythagoras Theorem. It is unknown whether it is actually Pythagoras that came up with the concept. He certainly brought it into popular use though. Many people refer to him as the originator of modern maths.

Isaac Newton is the father of calculus. Need I say more here? As you may know, he also came up with a number of calculations that help us understand this world a little bit better. For example, he was one of the first to come up with an understanding of what gravity was all about. This was all down to mathematics.

Leonardo Fibonacci is credited with the creation of the Fibonacci sequence. This is something which plays a huge part in our day to day lives. Do you know what he is best known for? He was the one who introduced the Hindu-Arabic Number system to Europe which then became known to the whole world. This is the number system that we use nowadays (i.e. 0-9). Basically, he was the one that helped to eliminate Roman Numerals from Europe.

Renes Descartes is credited with the invention of the modern graph system. This means that he was the one responsible for coming up with the Y and the X axis. He also came up with ideas about how to plot various things onto the graph. As you can probably guess, this is something which is in use to this very day. It doesn't matter what business you work in, you are going to need to use graphs for analysis at some point. The theories that he came up with are relatively unchanged, unlike most of the other people on this list.

Eular is considered one of the greatest mathematicians to have ever lived. It is believed that he came up with most of the mathematics theories that we know and love in this day and age. Very few of them are credited to him though. Some things that he helped to develop include number theory, graph theory, calculus and so much more.

Indian mathematicians

People are incredibly surprised when they find out just how much India has contributed to the world of mathematics. In fact, Indian Mathematicians have contributed so much to mathematics that without their work maths as we know it today would not exist. In this chapter, you are going to discover a little bit about some of the most popular of the Indian Mathematicians.

Aryabhata was one of the first people in the world to blend astronomy and mathematics into one. He was also one of the first people in the world to carry out work on approximating the value of Pi. The work that he carried out was still in use many hundreds of years later when an equation to start to find the value of Pi (we will never know the real true value of Pi) was invented.

Yativrsabha (also known as Jadivasaha) who lived around the same time (round 500-570) as Aryabhata pioneered using various units to help measure distances and time. He was also one of the first people in history to come up with the concept of infinity, or at least start to discuss it in his works. The book that he produced, known as the Tiloyapannatti was very popular among Indian Mathematicians around this time and it was still being used practically many hundreds of years later.

The last of the classical Indian Mathematicians which had an impact on the world of Mathematics was Sridhara who started to work out a formula that could be used to determine the volume of a sphere. His calculation wasn't 100% accurate, but it was accurate enough for the formula to be used in working on the actual calculation many hundreds of years later.

Indian Mathematicians have also had a huge impact on the way in which we deal with the number system in this day and age. For example, it was in India that the very first mention of zero as a number was recorded. It is believed that the Babylonians helped them work this concept out, but for the most part it was Indian Mathematicians that help spread the usage of the number and make it 'better' (it was in India that the first ever calculations using zero were carried out!). India also had a huge impact on the creation of the decimal system, which of course for most of us is very important when it comes to money and in calculations. The decimal system allows us to write numbers as large or as small as we want.

Perhaps the greatest achievement of Indian Mathematicians however was the development of trigonometry. This was developed by a number of top magicians at the "Kerala School of Mathematics and Astronomy" in the 1400s. They came up with the concept of Sine, Tangents and Cosine over 200 years before the rest of the world! Although of course the formula back then was not as advanced as the concepts that we have nowadays.

This is of course just a small sample of some of the Indian Mathematicians who had the most impact on the world. There are hundreds of others, particularly around the middle ages who made small contributions which helped to shape mathematics into the thing that we know and love today.

Ancient Indian mathematics

As you know already, most of what we know about the world of Mathematics was developed over 1600 years by the Indians. In this chapter, I want to take a little look at Ancient Indian mathematics in a bit more depth. This is the mathematics that started to be developed around 1200 BC.

The history of Indian mathematics is thought to begin in Prehistory. Various discoveries around Harappa (Indus Valley Civilization) have shown that it was in India that the concept of weights and measurements began to be developed. In 1200 BC, a fairly advanced system of weights and lengths was developed. There is also evidence that the people that lived around this time had a fairly decent understanding of basic geometry too! Evidence has shown that they were fairly accurate in standardizing lengths and weights, despite them not having the tools that we have nowadays to assist them with it. It was the ancient Indians that originally developed the idea of using a ruler to carry out measurements. Their one is known as the 'Mohenjo Daro Ruler'

Most of the maths that was developed in ancient India was 'invented' for use in religion. For example, it was during the Vedic Period that large numbers were used. These numbers appeared in texts such as the Yajurvedasamhita as 'mantras'. The numbers back then were given names. For example, a thousand was known as 'Sahasra' and 'Samudra' was the name for a billion.

Like the Pythagoras Theorem? Well, it was first mentioned verbally in Ancient Indian mathematics. Again, this was for the benefit of religion. The theory was used to help build altars in various temples. It is worth noting at this point that the Indians did not develop this theory. It was created by the Babylonians. However, the Indians were the first to write it down and begin to develop the idea of Pythagoras.

In 400 BCE the concept of 'infinity' began to be developed. Again, this was a concept developed in support of religion. One of the most interesting things about this time period is the fact that it was around this time that algebra began to be developed properly. Sure, there had been basic mentions of algebra throughout history. However, it was the

Indians who really began to develop the concept and come up with some new theories. This was also one of the first times that Algebra was really written down and solved.

Ancient Indian mathematics is also responsible for the development of 'common' concepts such as the decimal system (this is what we use today), the number zero, advances in the determination of the value of Pi and much more. In short, without the work that ancient Indian mathematicians carried out, maths would not be the same as it is today. As you can probably guess, it is very difficult to cover everything about Indian mathematics in such a short space. I therefore do urge you to carry out a bit of research yourself. You will be surprised at just how much you can learn about India and its mathematics.

Indian mathematics

If there is one form of mathematics that has shaped the world of maths more than any other it is Indian mathematics. India has always been at the forefront of mathematics development. In fact, some of the most important mathematicians in history have originated in India. This includes Bhaskara II, Brahmagupta and of course Aryabhata. Most of the contributions that they made to the world have helped to shape the way in which we view mathematics. Let's take a little look shall we?

Most people do not realize this, but concepts such as using zero as a number, arithmetic, algebra and negative numbers all originated from India. It is hard to imagine maths without them - isn't it? It was from India that they began to spread around the world. In fact, India helped to invent the decimal number system that we use today. They even had a hand in developing trigonometry. In fact, the more advanced concepts of it, such as sine and cosine came into in India by none other than Aryabhata himself and even documented in Aryabhatiya and Surya Siddhanta. It is thought that the word sine have originated from the Latin mistranslation of the Arabic word 'jiba', considered to be transliteration of the Sanskrit word for half the chord 'jyaardha'.

As you can probably guess, the history of Indian mathematics is rich. It is therefore virtually impossible to cover everything. I do want to go over a few highlights for you though.

One of the intriguing things about Indian mathematics is that it can be traced back a long way. In fact, all the way back to prehistory. Archaeological evidence unearthed from around this time shows that it was the Indians who were the first in the world to develop both a standardized system of weight and length. The system of course wasn't very sophisticated at that time, but it certainly was something quite Advanced.

Over the centuries, a variety of people contributed to the development of mathematics in the country. Most of what we know actually developed from religious texts. For example, a great deal of knowledge about geometry can be found in the 'Sulbasutras' dated from around 800 BCE to 200 CE. This is basically an appendix to religious texts namely the Vedas for use in construction of fire altars. The information here was used to construct altars. It was not really intended for the development of mathematics as a whole. However, this is how most forms of maths developed. It isn't just for 'fun'. It is to provide a practical solution to something. This is what the Indians have done throughout their history and invented things.

One of the things about Indian mathematics is that it has helped shape the world. Today inventors take their inventions for granted. The concepts that they came up with over the centuries were incredibly advanced for their time. They shaped everything that we do with maths nowadays. I mean, who could do anything without the numbers 0-9? Every single number in the world is made up of a combination of those 10 numbers. Everything that they taught us seems ever so simple now. As a result their contributions have gone a little bit unnoticed by many people since what they created are taken for granted and often not given any credit. I do urge you to take a deeper look into the idea of Indian mathematics. You will be surprised at just how much they taught us.

Introduction to Aryabhata

When it comes to Indian mathematics, particularly with regards to when it is used with Astronomy, there is no man more important to the development than that of Aryabhata. This man began to work on his concepts at the young age of 23. He then went on to play an absolutely massive role in shaping the concepts that we use today. Here, I want to give you a brief overview on what this amazing man was all about. We will go a little bit more into depth on the things that he did later on.

The work that brought the man to the forefront of Indian mathematics at just the age of 23 was the **Aryabhatiya**. The Aryabhatiya consisted of 4 sections namely; Dasagitika (10 verses), Ganitapada (33 verses), Kala-Kriyapada (25 verses) and Golapada (50 verses). These sections covered mathematics, time-reckoning and astronomy.

The concepts that the man began to develop in this written work were used for many centuries by Mathematicians all over the world. This book was developed in 'verse form' as a way in which to memorize some pretty complex things relating to the world of maths. A lot was covered in here including a section known as **'Kala-kriyapada'** consisting of 25 verses. This section covered a lot about astronomy, including how to determine where various planets were at certain times. This was the birth of 'Mathematical Astronomy'. It is worth noting here that Aryabhata was the first person (and the only person at that time) to develop an equation which could measure the radius of the Earth.

Aryabhata was also responsible for the development of a number of other major concepts that we use in mathematics nowadays. We will a closer look at some of these later on.

Concepts that he covered however included:

Developing a 'place value' system. Contrary to popular belief, Aryabhata did not actually develop the concept of using **zero**. However, his system pretty much implied that there was a zero there. Therefore he is often credited for its development.

He worked pretty hard on working out that Pi is most likely going to be a rational number. The formulas that he came up with to develop the **value of Pi** were used for many centuries after.

In addition to this, Aryabhata put a lot of work into the development of Astronomy. Much of what he said (such as insisting that the earth rotates daily) was disregarded at the time. However, we now know that most of what he said was true. In fact, like maths, most of what Aryabhata discussed about astronomy is still in use today.

The legacy of Aryabhata's work still echoes through our lives to this very day. Without the effort that he put into developing maths the number system that we use today would simply not exist. Over the next few chapters we are going to take a little look at the work that Aryabhata carried out in slightly more depth. We will also take a little look at the history of this fantastic man. You will be surprised at just how much he has done.

Name and birth place of Aryabhata

As you know, Aryabhata is perhaps one of the most famous Indian mathematicians out there. This man is responsible for developing many concepts that has shaped the way in which we view mathematics in this day and age. In this chapter, I want to discuss with you a little bit about the background of the man. In further chapters, we will take a look at the work that he carried out in more depth.

It is worth clearing up at the start that the man's name is Aryabhata. Throughout history, he is often referred to as Aryabhatta with a double t in "Bhatta". It is unknown why it was spelled this way in various texts. However the 'Bhatta' extension of the name does not really make sense. Therefore it is commonly believed to be just 'Aryabhata'.

It is unknown where Aryabhata was actually born. Some historian and mathematicians are of the view that he was born in Ashmaka, India. This would be in either the Deccan region of India or in the state of Kerala (Southern India). He was most likely born in 476 A.D. We know this because he mentions in his book (Aryabhatiya) that the year was '3,600 years into the Kali Yuga' and that he was twenty three at the time. Some say his native place was Kerala but perhaps never went there while yet others say he from Bihar where he was born and later worked at Nalanda University. In fact, there is a tradition in Kerala that Aryabhata and his son Devarajan was expelled from his community (some speculate Namputiri or the Nambudiri Brahmins) of 'castes' because they used to go to the sea very often and do observations. Perhaps the people used to think that they are bringing the wrath of natural elements. Most probably they were observing eclipses. It is possible that he was from Bihar but after the invasions of North India by outsiders, the centre of education shifted to Kerala.

Aryabhata came from a time period when very little was written down about anything. As you can probably guess, this means that there is a lot that we do not know about his life. It is unlikely that we will ever know more about him either. It is possible that he might have gone to various parts of the country including Kerala to carry out observations where his work was documented and hence the claim that he was from Kerala itself.

Ruins of Kumhrar in ancient Patliputra (previously Kusumpura)

When it comes to education, it is believed that Aryabhata received it in Kusumapura. In modern times this area is known as Patna. Based on his statement in Aryabhatiya, he most likely lived in this area at the time too. It is not actually written down anywhere that he received his education at this location. However, if you read through his book it is briefly touched upon that Aryabhata was the head of an institution around this area. It is believed that he was in charge of the

Nalanda University. There is no evidence to support this though. Around this time, it is also believed that he set up an observatory to carry out his astronomy work. It is believed that this was located at the Sun Temple in Bihar.

Ancient ruins of Nalanda University

As you may know, Aryabhata wrote a book known as the Aryabhatiya. You are going to find out a little bit more about this later. This book is responsible for shaping the way in which the world viewed mathematics for many centuries. He wrote it when he was just 23. This book went into depth on some very advanced concepts for the time including arithmetic, trigonometry, algebra, fractions and sines. This was coupled with discussion on astronomy too. Apart from this, the majority of the work that Aryabhata produced around this time was lost to history. This is a great shame because this may have set the world of mathematics back a good few centuries.

Much of the work that Aryabhata carried out is still reflected in the things that we know today. This includes the concept of Pi and even using Zero as a number. We shall later take a look at the works of this genius?

Taregna – The (birth) place of Aryabhata

Just nineteen miles away from Patna lies the small town of Taregna (Taregana). This beautiful town is home to less than 1,000 people. Most people would pass by it without a care in the world probably keen to explore some of the other fantastic areas of Bihar in India. However, most people do not realize that this is the birthplace of modern mathematics.

Taregna, a small place in Bihar

It is believed that Taregna was the birthplace of Aryabhata, one of the most forward thinking people in the world of mathematics. Even if he was not born here, the town still played a massive role in his life.

It was in Taregna that Aryabhata set up his observatory in the 6th Century. This observatory was located in the Sun Temple. It was here that he carried out a lot of research into the way in which our planet interacts with the rest of the universe. It is believed that he discovered the Heliocentric Model whilst working at his observatory. This has been disputed though. What is known however is that this was the FIRST place in the world that it was proposed that the Earth actually revolved around the sun. Of course, it took many years for people to actually believe this. However, Aryabhata was the first person to even entertain this idea and it all happened in this very sleepy town in the state of Bihar in North India. Today most people of Bihar let alone India would even know about this place.

It is believed that Aryabhata decided to set up his observatory here due to the fact that you got a pretty decent view of the sky. In fact, people realized this all the way up until 2009. When the solar eclipse rolled around, it was believed that Taregna was the best place in India to actually get a glimpse of it. As a result, thousands of people descended upon the small town. SA.D.ly though, clouds hampered the view in the end. However this does give us an indication as to why this location was chosen for use as an observatory. Of course, it was pretty close to his home town too, which I am sure played a major part in the decision.

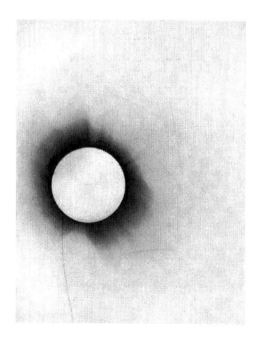

The town today pays very little attention to the work that Aryabhata carried out here. Although to be honest, it is such a small town that there is very little need to. The town is growing at a very fast pace right now. If you ever head to Taregna in Bihar, then you will be able to visit the actual location where Aryabhata had his observatory and made some pretty interesting discoveries. As you can probably guess, the buildings around here are nothing more than ruins now (although still just as beautiful, but seeing them fall into a state of disrepair like this without anybody really paying attention to their upkeep is a little bit sad considering the importance of the location) It is still a pretty exciting visit though for anybody that is interested in the world of modern mathematics.

The works of Aryabhata

Aryabhata is perhaps one of the most important people out there when it comes to the development of mathematics. The works of Aryabhata shaped mathematicians for centuries. It could be argued that without many of the concepts that he developed, much of what we know today could not possibly exist. In this chapter, I want to take a little look at some of the works of Aryabhata. As you may know, much of what he produced has been lost to history, although some very important things still survive.

One of the most important works produced by Aryabhata was the book 'Aryabhatiya'. He wrote this book at the age of 23. This book was written in verse form. It was designed to be an easy way to remember very complex things. Now, much of the work that was written down in Aryabhatiya was not new. However, Aryabhata was one of the first to write these concepts down and his book was used for many centuries by mathematicians. We will take a look at this book in more depth in later chapters.

One of the things that Aryabhata is credited for is the development of the number zero (0). It is worth mentioning though that he did not actually invent the number zero. This came much later. What he did invent however was a 'place value' system. This basically put a value to numbers based on their position. One of the reasons as to why Aryabhata is credited with the creation of zero is because the system he developed actually included hallmarks of using zero. He just didn't go out and use it directly!

Much of the work that Aryabhata carried out was the development of the approximation of Pi. He did not come up with an actual answer. However, some of the things that he said are still in use today. For example, he was the first person in history to state that Pi was an irrational number. He was also one of the only people that managed to come up with a rough formula for working out Pi at that time. What he wrote down isn't completely accurate, but it is close enough based on the tools he had at his disposal in those ancient times.

He was the first one to show how to solve indeterminate equations of the first degree and started to work on quadratic equations. In addition to this, Aryabhata made huge advancements in the world of algebra and trigonometry. Although just like the number zero, he did not come up with the concept. He did however go into a lot more depth than other mathematicians did on the subject.

In addition to Mathematics, Aryabhata put a lot of study into Astronomy.

He came up with theories surrounding the movement of the solar system and the way in which eclipses work. Much of what he said was dismissed at the time. However, we now know that most of the stuff that he said around this time is now true, particularly in regard to the way in which the Earth rotates on a daily basis. He is also credited with coming up with the heliocentric model of our Solar System.

The Arya-Siddhanta

Arya-Siddhanta is one of two popular sets of works that has been attributed to Aryabhata. The key difference with this piece of work is that we do not have full copies of Arya-Siddhanta. In fact, much of what the Arya-Siddhanta was about has not been found. But it is worth noting that not everything about the Arya-Siddhanta has been lost. As you probably know already, Aryabhata was one of the most popular mathematicians around that time. As a result, his work was readily shared amongst his contemporaries. They always referenced his work when carrying out their own. Therefore some of the knowledge from the Arya-Siddhanta comes from other mathematicians. Perhaps the most influential of these is mathematicians like Varahamihira, Brahmagupta and Bhaskara I. You will find mentions of this work pepper-potted throughout history.

It is believed that the Arya-Siddhanta was very focused on astronomy. His previous works were very mathematics influenced, but this was pure astronomy. Many of the concepts that were discussed in this book came from his previous works the Aryabhatiya (arguably his most famous of works) and the Surya-Siddhanta the oldest tradition archeoastronomy of the Hindus.

It is believed that the Arya-Siddhanta was an expansion of his work 'Surya Siddhanta'. This is because most of what he discussed was similar to the Surya Siddhanta, just slightly more in depth. For example, it was in that book that showed the midnight-day reckoning. In his works, the Aryabhatiya looked at it from sunrise.

In the book was a discussion about various astronomical instruments that were used by Aryabhata. It is unknown what most of these devices actually did and sadly that has been lost to history. Some of the things that we know about include:

The Gnomon: A Shadow Instrument (presumably for measuring the shadows to help determine the position of various celestial bodies).

Various instruments that could be used to measure various angles. Some of the ideas that he came up with in this book were developed in recent times for use in astronomy today.

He even used and devised his own devices such as various water clocks for measuring the time of various celestial bodies passing by. These were wildly accurate clocks. As you know, he came pretty close (within fractions of a seconds in fact) of determining the length of the day.

It is also believed that a lot of what Aryabhata discovered about astronomy over the years was written down into this book. Some of which include; the length of days, and perhaps intriguingly for that time, the fact that the Earth wasn't actually the centre of anything. He also came up with the fact that it was the earth that revolved around the Sun. As you have already found out this is something that was not believed at the time. In fact, Aryabhata was the first person in history that ever wrote this concept down. Most of what we know about astronomy nowadays stems from the work that comes from this book, even if it does not really exist in the original form that it was produced in. Aryabhata and his works remain incredibly influential to this day.

Who invented Pi?

Every single child out there dreads the time when they need to start learning about Pi at school. I won't lie to you. For some people it can be quite a boring subject. However, many people do not realize the rich history behind the number. Let's take a little look shall we?

For some reason, many schools out there teach the idea that a Swiss mathematician, who went by the name of Leonhard Euler came up with the idea of Pi. However it wasn't. Not by a long shot in fact. The first mention of the word Pi came a year before his birth. It was in 1706 that William Jones, a maths teacher, mentioned the concept in his book called 'Synopsis Palmariorum Matheseos', or loosely translated to 'A New Introduction to Maths'. This was a complete compilation of his teaching notes over the years.

However, when William Jones wrote about Pi it was not a new concept. In fact, it had been referred to as 'quantitas in quam cum multiflicetur diameter, proveniet circumferencia', which if you must know actually translates to what Pi actually does i.e. the number that the diameter is multiplied by in order to get the circumference. Basically, these people took the number quite literally. It is believed that William Jones was the first person to come up with the Pi symbol. As far as we know, he was actually the first person to write it down at least.

William Jones did have a major impact in the world of Pi beyond this too. Up until the writing of the book mentioned previously, Pi was always written as a ratio. This meant 22/7 or something similar. This meant that most people at the time believed that Pi was a rational number. We know now that it isn't. One of the concepts that Jones put forward (but ultimately never got around to proving) was the fact that Pi was an infinite number that would continue to repeat with no end in sight. We now know that this is true. However back then he obviously had no real way of testing it.

Jones was of course not the first person to use Pi to measure the diameter of a circle. In the previous century, the measurement was used by English mathematician William Oughtred who wrote a book which used Pi. However, the value of

Pi was determined by the diameter of the circle. Basically this meant that it always changed. This is very different to the Pi that we know (and love!) in this day and age.

Thousands of years before Jones was even born there were mathematicians looking to determine the value of Pi. Even now, we do not have a true value to Pi. There is no real inventor to the concept. It is just something that has evolved over time. It is Jones that set us on the pathway to our understanding in this day and age though, and we are closer to determining the value of Pi than we have ever been.

However not many education institutes say that it was Aryabhata who estimated the Pi. According to Walter Eugene Clark the author of the book "The Aryabhatiya Of Aryabhata", he says that it was Aryabhata who estimated the value of 'Pi' as: "Add four to one hundred, multiply by eight and then Add sixty-two thousand. The result is approximately the circumference of a circle of diameter twenty thousand. By this rule the relation of the circumference to diameter is given." In other words, Pi=62832/20000 = 3.1416, correct to four rounded-off decimal places.

The exact words of Aryabhata were:

"chaturA.D.hikam satamasaguam dvasasistatatha sahasraam AyutA.D.vayavi kambhasyasanno vrîttapariaha".

This translates to "Add 4 to 100, multiply by 8 and then Add 62,000. By this rule the circumference of a circle of diameter 20,000 can be approached."

This means in other words; Pi = 62832/20000 = 3.1416, correct to four digits.

Approximation of Pi by others and Aryabhata

Pi may seem like a rather simple number to us nowadays. What many people do not realize though is that many mathematicians battled for centuries to even come close to an approximation of the number. In this chapter, I want to take a little look at the development of Pi. I will not be able to cover absolutely everything in depth. This should give you an idea of the amount of effort that went into determining this number.

It is believed that the first approximation of Pi took place back in Egypt in their constructions. One of the best pieces of evidence of this is the Great Pyramid of Giza which incorporates some of the basics of Pi into it. There is no evidence that the Egyptians actually had a value of Pi. However, it is clear that they actually considered the idea. It is believed that the idea of Pi came from the Babylonians who brought it to Egypt.

One of the first people to actually make a push for the approximation of Pi was none other than Aryabhata. The work that he carried out in

499 A.D. was used for centuries afterwards by mathematicians. The work that he carried out was used to determine the circumference of Earth. It is worth noting that his approximation of Pi was not completely accurate. He got to 3.1416. I am sure you can agree that he got pretty close though, despite not having the tools available to him that we do have nowadays. The main thing that we can take from Aryabhata is that he discovered that Pi was an irrational number. This is something that we now know to be true.

Perhaps the greatest Advances in the approximation of Pi took place between the 16th and 19th Centuries. Quite a few people worked on an equation back then. Perhaps the best known was the German mathematician who went by the name of Ludolph van Ceulen. He calculated Pi to 35 decimal places. This was a pretty big accomplishment at that time.

The next big advancement came from Jurij Vega who managed to calculate Pi to over 140 decimal places. He was not completely accurate however. Only the first 126 numbers were correct. This was still a pretty big advancement though. The next person to come up with an approximation of Pi was William Rutherford over 52 years later. He managed to calculate it to 208 decimal places. Only 152 of these were right. The modern method that we use to determine the value of Pi today is based on the method that William Rutherford came up with.

The last 'large' advancement in the world of Pi came from Srinivasa Ramaujan, an Indian mathematician in 1910. He came up with a rather sophisticated formula to work out Pi which got us pretty close. The world of mathematics has advanced drastically since then though and we now have computers to work the value of Pi out for us. Currently we are up to 2.6 trillion digits, and that is growing by the day.

Aryabhata was not the first to use zero

In this day and age we take the number 0 for granted. It is absolutely vital to our numbering system. After all, removing a 0 from a long number will completely change the value all together. In the past however there was no usage of the number 0. This made equations incredibly complex. In this chapter we are going to take a little look at the first usage of the number zero.

It is worth noting that humans have always understood the value of zero....they just didn't have a number of it. As mentioned previously, this made making calculations fairly difficult. When the number zero was introduced, it completely changed the way that we view arithmetic.

To understand the first use of zero, we need to trace the history of the number back to when it wasn't a number. The first counting system in the world was developed by the Sumerians over 5,000 years ago. They used a system (believed to be angled wedges) which was positioned in a certain way. The position of this wedge determined what the value of it was.

It was the Indians that first developed the concept of the number zero. It is unknown how they came up with the idea. Some people believe that it was the Babylonians that passed on the knowledge, whilst others believe that it was completely the doing of the Indians. In either case, most history books claim that it was the Indians that came up with it on their own and that is the view that we are going to stick with!

It was in 458 A.D. that the concept of zero started to appear in India. At this time maths generally wasn't written down. Instead maths was done through chants and poetry. This meant no symbols. The words used for zero were generally 'space', 'void' and 'sky'. It was in 628 that the symbol for zero first came into existence.

It came from a mathematician and astronomer known as **Brahmagupta**. When he started using zero, it was a single dot under a number. However, he pioneered the usage of maths operations that used zero. This was something that had never been done before. Mathematical equations using zero nowadays are nothing special to us. Back then though this concept was absolutely ground-breaking.

Usage of the number zero very quickly spread around the world soon after this. When the concept of zero reached Baghdad, it was then that it became part of the Arabic number system. This is the exact system that we use in this day and age. It was the Persian mathematician known as Mohammed Ibn-Musa al-Khowarizmi who turned the number to the "0" that we all know and use today. The number 0 then travelled to Europe and it was slowly developed until it entered popular usage in the 1600s. This means that the zero arrived in Europe very late.

The real story of zero

Let me put it this way, "you think you know everything - just because you are "educated" and are well armed with all the scientific knowledge of the western world. Let me tell you something - you know anything. Well the answer is, you don't know really know anything. Most people will be quiet offended by this remark but what I am saying is, do you really know "zero". I am sure most of you know more than zero and less than zero but not the real story of zero itself.

In this ever dynamic world who cares a damn about this sweet little zero, but imagine if this is taken out of our lives all the scientific developments will come to a chaotic end and your four to eight figure salary will be a single digit trauma, NASA will have to cancel trips to mars and moon, our modern mathematics will be equally shaken and will find new abrupt faces to explain its complex theorems, in short the entire world will be a bear's garden in the absence of this single numeral called 'Zero' which is written as '0'.

By now you must have understood the impact of zero in your lives but still we are zero as most of them don't even know who discovered it. Unfortunately, for many Indians and in the west, the answer will be scanned among great scientist like Einstein, Galileo or perhaps Newton. These literates are too inspired by the awe of the western world and have forgotten that science and mathematics also existed even in the most ancient cultures and countries such as of Egypt, China or India.

Zero was not the brainchild of western world but the product of an Indian mathematician called Brahamagupta in 598 A.D.. Brahamagupta was born in Gujarat in the city of Bhinmal which is presently in North West Rajasthan. Brahamagupta was the head of the department of mathematics and astronomy at the astronomical observatory Ujjain University which was at that time a great centre for learning the science of prediction called astronomy.

In chapter 18 of his famous book called Brahmasphutasiddhanta (Corrected Treatise of Brahma) Brahamagupta describes about zero as one of the numerals which stood for meaning nothing. He also elaborates as to how integers positive and negative consequence when played with zero.

For example:

Addition with zero - If a positive integer value is added to a similar negative value the result is zero, the sum of a negative integer and a zero is negative, sum of a positive integer and a zero is positive and the Addition of two zero's is zero.

Subtraction with zero - A negative integer subtracted from zero is negative, a positive integer subtracted from zero is positive and a zero subtracted from zero is zero.

Multiplication with zero - A negative integer, positive integer and a zero with Zero will always be Zero.

Square and square root of zero - The square of a zero and square root of a zero is always zero.

Division with zero - A zero divided by a zero is zero and a negative or positive integer divided by zero is zero.

Only at this point of division Brahamagupta makes crucial mistake as we all know that an integer divided by zero is not zero but infinite.

Brahamagupta is criticized by many contemporary and ancient scholars and mathematicians for this ardent mistake but what they fail to realize is the very foundation Brahamagupta provided to the platform of science and mathematics by using Zero as numerals for the first time and that too successfully. There is another side of the coin which still needs to be reviewed. It states that an integer when divided by 0 is 0 which most of the scientists believe is wrong but the synonym of 0 is "shunaya" which is fortunately also the same for "infinite". However it is a matter of debate and still in hypothesis whether Brahamagupta knew about the concept of infinite/undefined or not. But scanning at his astronomical geniuses, it seems that he did know it and he was well acquainted with the theorem.

Zero is not just a word or a numeral it is a symbol of the pride of the East and the fact that the eastern culture's research and development in the field of science and mathematics is the very foundation on which the medieval western discoveries were made and yet more future ones still remain to be made through forgotten zero of the east.

History of algebra

Out of all of the mathematics concepts, it is perhaps Algebra that has the richest history. There are two types of Algebra around at the moment. The most common is 'classical algebra'. This is where you are tasked with finding out what an unknown number is. I am sure that you have done these equations in school before. This type of algebra has been developed over 4,000 years by mathematicians. Let's take a little look at a rough history of it shall we?

The first known usage of Algebra was in Ancient Egypt. The first mention of algebra was found on a document now known as the 'Rhind Papyrus'. It was created in 1650 BC. The problems here were linear equations and the Egyptians were able find one unknown verbally using the equations. If we fast forward to 300 B.C. then a second Papyrus was written. This is known as the 'Cairo Papyrus'. It was clear that by this time the Egyptian's understanding of Algebra had evolved quite quickly as they could now work out two unknowns. However their maths was incredibly cumbersome and thus they would not have been able to improve much on that.

At the same time as the development of the Rhind Papyrus, algebra was also being practiced in other parts of the world. This was by the Babylonians. It was clear at this point that they were a great deal more Advanced in Algebra than their Egyptian Counterparts. Like the Egyptians they solved their problems verbally and they did not use symbols in their work. The Greeks at the same time worked on Algebra. This was similar to Babylonian Algebra, although the Greeks did not recognize irrational numbers. This meant that they referred to quantities as geometrical magnitudes instead. One thing the Greeks did give us was the power of deductive reasoning when it came to solving algebraic problems something which the previous mathematicians did not do.

Over time the concept of Algebra began to develop all over the world. A number of new concepts were introduced. For example, the Hindus and Arabs were keen to use irrational numbers in their equations, which made everything a great deal simpler. The various forms of Algebra began to merge in around 1050 A.D. when it was much easier for people to travel between countries.

It was in the 1500s and in Europe when Algebra began to be developed quite drastically. This was around the time that zero was now considered a number and irrational numbers were commonplace. In the 16th century a number of concepts were introduced. This included cubic and quartic equations. Most people regard this period as the start of modern mathematics due to the amount of advancements that were made.

Symbols were introduced in France by Viete. This really helped to take Algebra to a whole new level, and it was much easier for many people to study the relationship between numbers in these equations.

Finally, in the 19th Century the British began to develop a 'break off' form of Algebra known as Abstract Algebra. This isn't studied in schools quite as much. This form of algebra looked at mathematical objects such as vectors, transformations and matrices. This form of Algebra is being developed quite rapidly at the moment.

Today, Indian algebra is called Modus Indorum" or the method of the Indians. The algebra that was created by the Indians was propagated by the Arabs which then went to Europe. Today, many concepts used in Algebra to solve equations was first used by Aryabhata. Some of this other works were taken by others and expanded.

Aryabhata and algebra

As you know, Aryabhata made a number of advances in the world of mathematics. One of the most important contributions he made was to Algebra. In fact, much of what he produced was the first time that things had ever been written down related to the subject. He also produced the concepts in a fairly easy to understand way. This was important for helping mathematicians to get to grips with his concepts and advance them even further.

Let's take a little look at the advancements that he made shall we?

It is worth noting that much of what Aryabhata came up with when it comes to Algebra had routes in his astronomy. It was not to carry out more research into the world of mathematics. He was an expert here. What he saw was Algebra as a way to give him tools to work out the position of planets. The equations that he came up with were used to understand speed and velocity etc.

I think one of the most exciting (and important) concepts that Aryabhata came up with is the **'Problem of the Messengers'**. This, as you may have guessed was related to his astronomy. This was a calculation that was discussed all over the world at the time. It basically was going to be used to determine when two planets, which moved in opposite directions, or in the same directions met.

One of the key points to note about the algebra that Aryabhata came up with was the fact that they were formulated in a way to avoid negative numbers. This was changed later on so we do not have the equations he came up with exactly. For the most part we have them in place though. This is what he came up with (the English translation quoted in many different books since that point):

"Divide the distance between the two bodies moving in the opposite directions by the sum of their speeds, and the distance between the two bodies moving in the same direction by the difference of their speeds; the two quotients will give the time

elapsed since the two bodies met or to elapse before they will meet"

In the Aryabhatiya, he provides what he calculated about the summation of series of **squares and cubes**.

$$1^2 + 2^2 + \ldots + n^2 = n(n+1)(2n+1)/6$$
$$1^3 + 2^3 + \ldots + n^3 = n(1+2+\ldots+n)^2$$

Most of what he had done however had been discussed previously for many centuries. It was Aryabhata who was the first to write these concepts down though. Much of it was put into a simple to read form as a way in which to aid memory in mathematicians. A lot of this work actually survives and gives us a great deal of understanding about this time period.

This is just a small amount of the things that Aryabhata produced over the years. As you know from the rest of this book, a lot of the work that he produced has been lost. Therefore we do not know everything that he came up with. Regardless, what we do know about was used for many centuries afterwards. In fact, much of what he taught us you have probably learned in schools, or at least in modified versions of it.

Aryabhata and trigonometry

Whilst Aryabhata is best known for his advances elsewhere in mathematics, for example, working to determine the value of Pi and other functions, he is also known as one of the 'forefathers' of modern trigonometry. It is worth noting that he did not actually invent the concept. Instead, he helped forward the topic a lot. Much of the work that he carried out when it comes to trigonometry is still in use today. In fact, you may have even learned some of the things he accomplished whilst at school.

Let's take a little look at the basics of what Aryabhata came up with when it came to trigonometry? As mentioned previously, most of what is discussed here had been carried out elsewhere in history. However, it was Aryabhata which first gave definition to the various terms. Most of these definitions were written down in various works that he published over the years. Some of these works have been mentioned elsewhere in this book. As you know, a lot of what he wrote was still in use for many centuries afterwards.

One of the most important advances Aryabhata made was coming up with various **'basic trigonometry functions'**. Some of these functions were written down for the very first time. They were expanded numerous times over the following years. However it was Aryabhata that first gave definition to them.

One of the main things that we discuss today in mathematics at schools is determining the **'area of a triangle'**. This idea stems from some of the things that Aryabhata wrote down. In one of his most popular works he wrote this:

"for a triangle, the result of a perpendicular with the half-side is the area"

Aryabhata continued to make advances in the world of trigonometry. One of the most important things he done was define what **'sine'** actually was. This is a definition that we use today. Basically he was the first person to determine that it was the **relationship between half a chord and half an angle**. He determined this not from his mathematical observations but his astronomical observations. In addition to this he is credited with giving definition to:

Cosine

Versine

Inverse Sine

He also started to come up with a number of calculations which could be used to determine the approximate value of each. He was not 100% accurate here, but the work that he carried out was developed over the years and he was pretty close in his approximation. He was one of the first people in history to **draw up tables** to work out the values of each of these three concepts.

As you can see, Aryabhata contributed a lot when it came to trigonometry. The amazing thing is that this is only the work that we know about. Much of what he produced was lost. It would be interesting to see just how many other concepts he came up with related to the subject that we will most likely never find out about. Who knows where we would be if we had access to his knowledge that we've lost forever.

Indian astronomy and Aryabhata

Much of what we know about astronomy comes from India. In fact, many of the world's greatest astronomers heralded from here. In this chapter, I want to give you a brief overview of the way in which Indian astronomy evolved over the years.

The earliest references to astronomy in India can be found in a set of books known as the Vedas. These texts, as you can probably guess, were very religious in nature. The books were formed of series of hymns which were put together over hundreds of years. These hymns literally changed the way in which the Indians saw the sky, and they are credited with many of the advancements that we saw in Astronomy around that time, even if they were not 100% accurate. These hymns talked about the stars in the sky and the way in which the sun rose every single day.

As time progressed the way in which the Indians viewed astronomy changed significantly. No longer was the view of the sky spiritual. It became a great deal more scientific. Around this time people in India began to believe that the sun was a star, just like everything else. One of the interesting things about this time period was that the Indians were the FIRST people to believe that the planet was shaped like a sphere. Even those in the Western Hemisphere still believed that it was flat at this time. A fantastic Indian astronomer, who went by the name of Aryabhata (see how his name pops up again), was the FIRST person in the world to ever mention the idea that the **sun was at the centre of our Solar system** and it was **us that revolved around it** not the other way around. Again, western astronomy had not caught up to this point.

Other discoveries that Aryabhata made around this time included estimating the way in which the **planets and stars moved around**. He was able to measure **how long it took the earth to spin on its axis** to within a fraction of a second. He also was three minutes off measuring **how long a year lasted**. This was pretty advanced for his time due to very limited methods of calculation. It is surprising to many people now just how accurate he was.

Despite the 'discover' of gravity being attributed to Isaac Newton, it is believed that many Indian astronomers started to dabble with the theory of **Gravity in the 6th Century**. It is unknown how far they took this idea though. Like most other ideas though, this was purely spiritual. They believed that spiritual bodies were held in place by some force. They were undecided what actually held them in place though.

After these few discoveries the world began to become smaller. This meant that idea began to be shared around the world. Most of what we know today as a result has some root in Indian astronomy. It is just a great shame that most people forget about the origins of astronomy.

Astronomical observations of Aryabhata

Aryabhata was one of the most 'forward thinking' astronomers in history. Many of the theories that he came up with, although dismissed at the time, are still used today. In this chapter, I want to take a little look at some of the Astronomy work that Aryabhata carried out at his observatory.

अनुलोमगतिनौंस्थः पश्यत्यचलं विलोमगं यद्वत् ।
अचलानि भानि तद्वत् समपश्चिमगानि लङ्कायाम् ॥

Mention of the rotation of the earth on its axis by Aryabhata

Perhaps one of the most important things that he looked into was the way in which the **solar system works and moves**. It is believed that he was one of the first people in history to come up with the idea that the **Earth revolves around the sun**. This is of course something that was rejected by most scholars for a while, but we do know that it is true. He also stated that the **Earth rotates once each and every day**. His theories also enabled him to work out where and when various planets would be located. This all derived from him studying the sky every single evening. It is believed that he came up with the **heliocentric model**, although this is disputed. Regardless, he was still one of the people that contributed to the basics of this model.

Other work that Aryabhata carried out was the study of **solar and lunar eclipses**. He stated that the **Moon and planets did not actually shine themselves**, something which previous theories had suggested. He stated that these planets only **reflected sunlight**. Again, this is something that we know to be true. He said that a **Lunar eclipse** (i.e. the full moon, half-moon etc.) occurred when the moon entered the shadow of the Earth and some of the sunlight was blocked. His calculations for working out about eclipses were improved later on by many other Indian mathematicians. He came up with the basics of the idea though. He was able to calculate to within a couple of seconds just **how long a lunar eclipse would last**. This is a pretty big accomplishment considering he had very little tools available to him at the time.

In addition to this Aryabhata looked into the idea of **Sidereal periods**. This was pretty much the **rotation of the Earth on its axis**. He determined this by looking at the stars in the sky. The method for calculating time back then was completely different to what we have now. If you look at it in modern terms though, he came up with the idea that the **rotation of the Earth** was 23 hours, 56 minutes and 4.1 seconds. He got extremely close (the actual time is 23 hours 56 minutes and 4.091 seconds). When it came to the **calculation of a year** he was just three minutes and 20 seconds off. That really is not bad when you consider the fact he worked everything out just by looking at the sky!

The ideas that he came up with related to **Sidereal periods** were used (and are still used today) by astronomers all over the world. Many of them have improved on his work, but they still use his theories as 'base'.

Heliocentrism and Aryabhata

Heliocentrism may seem like a pretty complex term. To be honest though, it is rather simple. All heliocentrism looks at is the idea that the Earth and other planets revolve around a stationary object i.e. the Sun. Let's take a little look at the theory in depth shall we?

As you may know, in the past, many people believed that it was the sun and pretty much every other planet that revolved around us. Basically people believed that we were at the centre of the solar system. It was not until the 16th Century that the Western world began to accept the idea that this theory might not quite be right. In fact, it was a theory that was wildly wrong.

The first discussion of the idea of heliocentrism came from Aristarchus of Samos back in 260 A.D. The work that he carried out has been lost, although we do have the observations of some of his contemporaries. He worked out that the sun was considerably bigger than the earth (although his writings put it at only 6 or 7 times the size). He did believe that this size difference is what made it more likely that the earth moved rather than the sun. He did not develop the idea much further than this though.

One of the first astronomers to continue to develop the idea of heliocentrism was Aryabhata, although the theory was not quite as developed back then. He was the first to notice that it was actually the earth (and other planets) which rotated around the sun. He came to this idea by observing the way in which the stars and other planets moved around the earth. By doing this it was clear that we could not possibly be right at the centre of the solar system. One of the things that the theories from Aryabhata lacked was an actual model of our solar system. Back then it would be pretty impossible to predict using the tools that he had at his disposal. It is only relatively recently that we have come to understand just how the solar system is put together.

One of the great things about Aryabhata was that he was able to accurately predict the orbit of every single planet. He knew where each planet would be and at what time. This was all central to the idea of heliocentrism. It is worth mentioning that his name was not proposed until long after Aryabhata's death.

As you can probably guess, one of the main reasons as to why heliocentrism was not readily accepted by most people was the fact that it conflicted with religion (which is pretty much why most ideas were rejected around that time!).

Gradually over time attitudes began to change, and this is the model that we know today.

References and further reading

Clark, Walter Eugene. 1930. The Aryabhatiya of Aryabhata: An Ancient Indian Work on Mathematics and Astronomy. University of Chicago Press; reprint: Kessinger Publishing.

Subhash Kak. 2005. Aryabhata and Aryabhatiya. Essays in Encyclopedia of India (edited by Stanley Wolpert). Charles Scribner's Sons/Gale, New York.

K Chandra Hari. Historical notes. Eclipse observed by Aryabhata in Kerala.

Anand M Sharan. 2007. Famous astronomer and mathematician Aryabhatt 1 of Kusumpura. Faculty of engineering memorial, University of Newfoundland.

Georges I. 2005. The Universal History of Numbers – II, Penguin Books India, New Delhi.

Encyclopaedia of the History of Science, Technology, and Medicine in Non-Western Cultures. Helaine Selin. Springer; 2nd ed. 2008 edition.

Ansari, SMR. 1977. Aryabhata I, His Life and His Contributions. Bulletin of the Astronomical Society of India 5(1):10.

Further information on Aryabhata can be obtained from http://aryabhatta.net

Made in the USA
Columbia, SC
25 January 2018